The Technical Writer's Guide

The Technical Writer's Guide

Robert McGraw

SkillPath Publications
Mission, KS

©1996 by SkillPath Publications, a division of The Graceland College Center for Professional Development and Lifelong Learning, Inc., 6900 Squibb Road, Mission, Kansas 66202. All rights reserved. No part of this publication may be reproduced, stored in a retrieval system, or transmitted in any form by any means, electronic, mechanical, photocopying, recording, or otherwise, without the written prior permission of SkillPath Publications.

Project Editor: Kelly Scanlon

Editor: Jane Doyle Guthrie

Page Layout and Cover Design: Rod Hankins

Library of Congress Catalog Card Number: 95-72618

ISBN: 1-57294-012-3

10 9 8 7 6 5 4 3 2 1   96 97 98 99 00

Printed in the United States of America

Contents

Introduction 1
1 Technical Communication 3
2 Preparing 17
3 Thinking 29
4 Writing 47
5 Editing 67
6 Proofreading 83
7 Some Common Types of Technical Writing 93
Conclusion 109
Appendix A: Quick-and-Easy Punctuation Guide 111
Appendix B: The Gunning Fog Index 123
Appendix C: Answers to Exercises 125
Bibliography and Suggested Reading 131

Introduction

Most people skip over the introduction in a book, but you're reading it. That means you already have the main criterion for being a good writer: You care about words. Writing is a method of recording words and thoughts in symbols so we can transmit them to another person at some future time. Many ancient cultures believed that the ability to change words into meaningful symbols was a gift from the gods. When you consider the enormous power of the written word throughout history, it's not hard to believe they were right. Yet for all its power and influence, writing itself isn't the goal. It's only a means to an end: communicating ideas.

This book is designed to help you sharpen your technical writing skills in a format that progresses just as the writing process itself does. There also are self-teaching exercises throughout the book. Don't skip over them: They really will help you. At the end of the book is a short, easy-to-use reference guide that will help you handle many of the most common problems in punctuation and grammar.

Above all, this book will teach you, remind you, and encourage you to be clear and concise. The English cleric Sydney Smith once wrote, "The writer does the most who gives his reader the most knowledge, and takes...the least time."

So don't waste any more time reading the introduction. Turn the page and start sharpening.

Chapter 1

Technical Communication

Although the business world commonly refers to writers who create manuals, draft proposals, and compile reports as technical *writers,* technical *communicators* would be more accurate—our challenge is not merely *to write,* but to *communicate.* In fact, the word *communicate* comes from a Latin word meaning "to share or to have something in common." The job of any writer is to share ideas with the reader. It doesn't matter how many writing classes a person has had or how much time and effort goes into writing the document; if the reader doesn't understand, then the writer has failed.

Here's the first of several good writing rules for technical writers:

Rule #1: Always keep your reader in mind.

The Process of Communication

Four simple questions should help you shape every document you write:

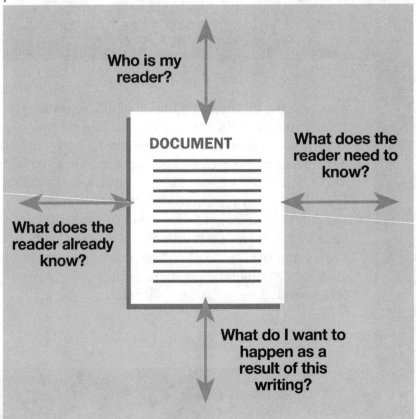

Look at these columns of words. What do they have in common?

A	B	C
encoder	symbol	decoder
sender	message	receiver
writer	idea	reader

Perhaps you noticed that all the words in column A refer to the same person: you. All the words in column B relate to the document you're writing, and all the words in column C describe your audience.

Instead of looking vertically, you can also look horizontally. Each horizontal row presents a different way of describing the communication process. When you write, you act as the sender/encoder by translating ideas into symbols on a page. The reader/receiver has the task of deciphering your squiggles and ending up with (you hope) some meaning fairly close to your original idea.

Every writer knows writing is hard work, but what about reading? If you want a real exercise in frustration, try reading a badly written instruction manual, such as the one that probably came with the last electronic device or software program you bought. The following example came from an instruction booklet for a camera:

Exposure Compensation Mode

When making your picture finished overall lighter, shooting a figure against the light, in a skiing ground, or photographing the shadow of a building under the very contrasty lighting to be brighter, this function is very effective.

Technical Communication

If the function is effective, that's more than can be said for the sentence. Copy down your own favorite bit of gobbledygook and think of it whenever you start to write. It will help you remember how difficult being a reader/decoder sometimes is.

Which brings up Rule #2:

Rule #2: Never forget Rule #1.

If you don't serve the reader's needs, you might as well hang up your keyboard and go into some other line of work.

Exercise #1:
Getting Your Point Across

To catch a quick glimpse of the challenge in communication and the instant impact of the words and phrasing you choose, try this exercise. You'll need a friend for this one, and the friend will need a pencil, paper, and three free minutes. Look at a photo or drawing without letting your friend see it; then put it away. Now describe to the "artist" what you saw in the picture. (No fair using gestures; describe only with words.) At the end of three minutes, compare your friend's drawing with the original image.

How well did you communicate the visual idea? Don't feel bad if the drawing isn't too close to the original. True, this isn't a writing exercise, but it's a good exercise for writers. It will help you appreciate the difficulty of translating ideas into symbols efficiently and the problems of decoding those symbols accurately.

The Role of Technical Writing

Technical writing is a form of factual writing in which the purpose is to explain things; that is, to translate technical and scientific information into terms a target audience (the reader) will understand. There are many kinds of technical writing:

Abstracts	Product descriptions
Articles	Progress reports
Feasibility reports	Proposals
Instruction manuals	Rules
Investigative reports	Scripts
Lab reports	Specifications
Memos	Test reports
Press releases	and on and on…
Procedures	

In fact, almost any writing done in the business world can be considered a form of technical writing. Nearly all business writing serves to document some decision, action, process, or result. When you write any kind of report or make a written record of something that happened, you're doing technical writing.

Good technical writing is clear, to the point, and objective. Bad technical writing, on the other hand, is fuzzy, overwritten, and lifeless or uninteresting.

There are four basic obstacles to good technical writing:

1. The writer doesn't understand the subject.

2. The writer understands the subject generally but can't distinguish the essential points from unnecessary details.

3. The writer understands the subject but is a poor communicator because of:

   - Foggy thinking.
   - Poor grammar.
   - Bad spelling.
   - Limited vocabulary.
   - Faulty usage.

4. The writer doesn't write appropriately for the target audience.

Writing to Communicate Versus Writing to Impress

The biggest problem in written communication is that too often writers forget that the goal is to share ideas with the reader. Instead, some writers allow another goal to take over: the desire to impress people. That's natural. All writers are subconsciously aware that whatever they write is going to become, to a certain degree, public and permanent. In 1975, archaeologists excavating the ancient city of Ebla in Syria found some of the oldest written samples in the world. Can you guess what most of the records were? That's right—day-to-day business reports. Accounts of who sold what to whom for how much and how it was shipped.

No one wants to leave something for posterity that might make them look less than intelligent or competent, so many people unconsciously become concerned with *how* they communicate rather than *what* they communicate, or even *if* they communicate at all. Wanting to impress people may be natural, but it doesn't help anyone communicate effectively. If those Eblaite clerks 4,500 years ago had worried about today's scholars critiquing their reports, the camels would still be standing around waiting for the next shipment. When it comes to technical writing, let posterity take care of itself. Forget about impressing people and just concentrate on getting the message across. The result will be shorter, clearer, more effective documents.

Signal-to-Noise Ratio

Audio and video engineers talk about signal-to-noise ratio, which means the amount of clear, usable electronic information getting through to the receiver versus the static and interference that degrade the sound and image. In writing as in broadcasting, signal-to-noise ratio must be kept to optimum levels. Some people write as though they think they're getting paid by the letter. The words are unnecessarily fancy, the sentences are complex, the paragraphs are convoluted. Often readers have to read the material two or three times before they finally understand what the writer means. The writer's signal-to-noise ratio is poor. If something isn't helping to transmit a clear message to your reader, leave it out. It's just noise.

Here are two examples of how people write when they really want to get their ideas across quickly and clearly:

- **To-do list:**

  Pick up enough low-fat milk to last the kids over the weekend.

- **Note to a co-worker:**

  Terry, Help! I have to give our illustrious leader a project update on Monday. We need this to be good, so please give me a copy of your fabulous charts and transparencies no later than noon tomorrow (my plane leaves at 4 p.m.). Thanks. I owe you one.

  —JD

Now picture those messages in the foggy, overwritten style writers so often use in business and technical writing:

- **To-do list:**

  A quantity of milk is to be obtained appropriate to the average needs of two children, aged six and eight, and one adolescent, aged twelve, and sufficient for a period of time not less than forty-eight (48) hours nor more than sixty (60) hours and containing a butterfat content of not less than one (1) nor more than two (2) percent by volume.

12

- **Note to a co-worker:**

    Terry, your immediate assistance is urgently needed and will be greatly appreciated. It has been requested that yours truly meet with the company's chief executive officer Monday next, at which time it will be expected that a formal presentation be given concerning the most recent relevant developments on the project.

    In order to adequately convey the importance of our enterprise, it is requested that you supply me with copies of the very excellent charts and transparencies which were produced by you. Please ensure that the above-referenced materials are in my possession no later than noon tomorrow, in order that I may take them with me when my airline flight leaves at four in the afternoon.

    Naturally, your cooperation will be considered a favor requiring a quid pro quo at some future date.

    Respectfully, your colleague,

    —JD

Bordering on the absurd, right? (Maybe even a few yards across the border.) So what happened? In both cases, the writing wasn't appropriate to the needs of the reader. The writer forgot Rule #1.

Have you forgotten Rule #1? See if you can write it in this box (from memory, please!):

Billboards: A Quick Course in Concise Communication

Someone has estimated that by the year 2000, the average executive will have to read one hundred forty pages of work-related material every day. So you can see, it's important to get to the point when you write.

Good writing is like a billboard: it's clear, it gets to the point, it can be understood quickly and easily. A billboard must grab the reader's attention and get a message across in three seconds. In order to do that, really effective billboard messages are no longer than seven words. Here are some examples you've probably seen:

Be all that you can be. (The U.S. Army)

Have you driven a Ford lately? (Ford Motor Co.)

Only you can prevent forest fires. (The Forest Service)

The Proud. The Few. (The U.S. Marine Corps)

McDonald's—Next exit.

These messages are short. They grab the reader's attention and plant an idea firmly in his or her mind. They communicate well and they communicate quickly. They work! If organizations are willing to pay ad agencies thousands of dollars to boil an idea down to seven words, maybe there's a good reason.

Rule #3: Don't overwrite: You're not being paid by the letter.

Technical Communication

Exercise #2:
Billboarding Your Message

How are you at creating a short, clear message? Write three or four billboard messages here, remembering that:

- The message must be no longer than seven words.
- Your reader must be able to read it and understand it in three seconds.

If you need a situation to get you started, try one of these: a line of premium pet food, a medical research program, low-interest home equity loans, a voter registration awareness campaign.

_____
_____
_____
_____
_____
_____
_____
_____
_____
_____
_____
_____

The Technical Writer's Guide

Every writing task can being divided into five levels:

- Preparing
- Thinking
- Writing
- Editing
- Proofreading

The next several chapters will correspond to these levels. Plus, as you explore each level, you'll also deal with "The Writer's Six Problems":

- Objective not defined
- Insufficient knowledge
- Thesis (main idea) not focused
- Audience not targeted
- Writer's block
- Organization not logical

These problems generally occur in level one (Preparing) and level two (Thinking). If you're well in control of the first two levels, you've already won half the battle.

Chapter 2

Preparing

Many successful authors put as much time into organizing and preparing as they spend actually writing. Novelist Danielle Steele, for example, spends up to a year planning a new book. Then she writes the first draft in ten to fourteen days and spends another two to five days rewriting.

Thorough preparation is critical to good technical writing, and there are several keys to successful preparation:

1. Identify your objective.
2. Know your topic.
3. Focus your ideas.
4. Target your audience.

Writer's Problem #1:
Objective Not Defined

Ask yourself, "What do I want to achieve? Why am I writing this piece? Should it instruct the reader? Convince? Inform? What do I want to happen?"

If you don't know what you want to happen, you can't expect your reader to fill in the blanks. For every document you write, you must have an objective clearly in mind. A good objective has three characteristics:

- Definable
- Achievable
- Measurable

Definable

You should be able to state, or better yet write down, what you want your document to accomplish. Here's an example:

> *This executive report will explain the steps our department took to streamline the field engineer reporting procedure. The readers (our regional management staff) will understand how we were able to reduce the number of forms from six to three, and cut the average length of the reporting cycle from five days to two.*

Achievable

A good objective is realistic and possible.

This procedure memo will take field engineers step by step through the process of filling out the AP27 form, and show them how completing the form at the work site can save them as much as forty-five minutes on each report. This will motivate them to take the forms with them rather than waiting to complete the paperwork back at the office.

Measurable

Where possible, devise some way that you can look back later and see if the document actually achieved your objective and did the job you wanted it to do.

Last year, before we wrote the new procedure memo, field engineers spent an average of 6.5 hours per month completing AP27 forms. Since the new memo was distributed, the average time has dropped to 5.1 hours per month.

Writer's Problem #2: Insufficient Knowledge

If you don't know enough to write about a topic, the solution is clear-cut: educate yourself.

While you're gathering information, keep one thing in mind: Although you need to understand the topic, you don't have to become the world's foremost authority before you can start writing. Don't fall into the trap of spending so much time doing research that you wait until the last minute to actually begin composing. Some people use the research phase as a kind of "justifiable writer's block."

No matter how much you know about a subject, it won't help anybody unless you can communicate that knowledge. Get the information you think you need and start writing. If necessary, go back later and do more research to strengthen your weak areas, but don't procrastinate—write. Working on the piece will quickly show you where your deficiencies lie.

Some of the worst technical writing is done by people who understand the subject inside and out. Expertise is not always the advantage it may seem; authorities often assume readers understand more than they actually do. Read some software "documentation" sometime. (That's what the computer world calls instruction manuals.) You'll quickly realize that being less than an expert can actually work in your favor. This perspective often will help you see what areas are likely to be confusing, what questions really need to be answered, and what should be the most logical approach.

Writer's Problem #3:
Thesis (Main Idea) Not Focused

To sharpen your focus, ask yourself:

- What information should I include?
- What should I leave out?
- What will help the reader understand the main idea quickly?

Sometimes it's difficult to decide what's necessary and what isn't. Think of a document as a body. The main objective and its supporting ideas make up the skeleton. The connecting transitions and clarifying details become the flesh and skin. Everything else is fat. Effective writing is lean, tight, and hard working, so it's important to trim off the fat.

To help decide what information is important, first get all your ideas in a rough form. A good technique for gathering ideas is called *web-charting* (some people also call this *clustering* or *mind-mapping*). The tools are simple—just a pencil and a piece of paper.

In the middle of the paper, write a key word or phrase to identify the topic of your document or the objective you want to achieve. Now, quickly write down every idea that pops into your head about that topic. Words or short phrases are fine. Don't be concerned about complete sentences, spelling, grammar, punctuation, form, structure, or any of the other etceteras. Don't worry about organization at this point either. Don't even think about whether your ideas are good ones—just write them down (later you'll go back and weed out). Having all these ideas down on paper will prove a great help in deciding what does or doesn't contribute to the process of clear communication.

When you're finished, your page will look something like this:

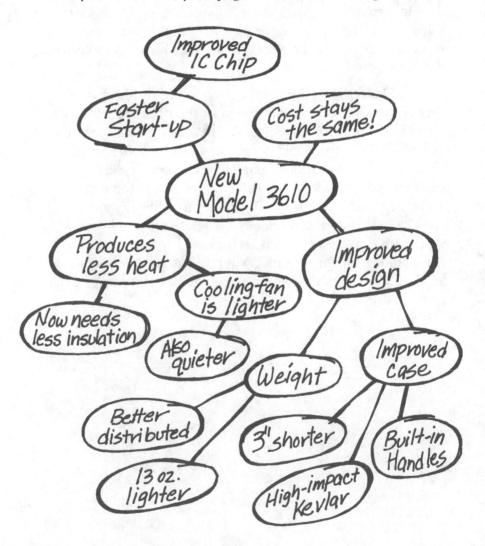

Or perhaps this:

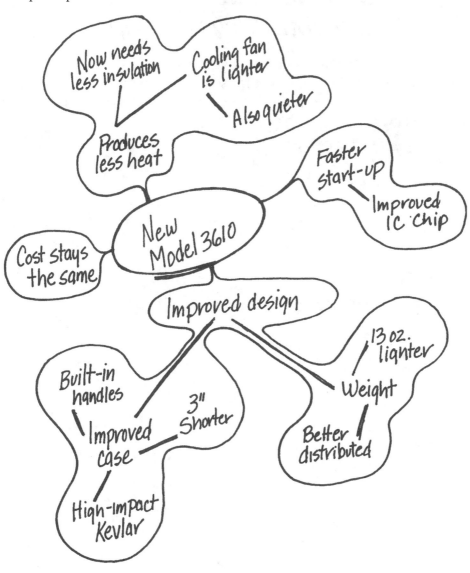

Exercise #3:
Getting Your Ideas on Paper

Take a few minutes now to give this method a try. In the space below, create a web-chart for something you may have to write. Remember, neatness doesn't count—just come up with as many ideas as possible.

Writer's Problem #4:
Audience Not Targeted

In order to understand a document, the reader doesn't need to know a thing about the writer. The right words are either on the page or they aren't. The meaning either comes through or it doesn't.

On the other hand, the writer should know as much as possible about his or her readers. Everything you can learn will help you do a better job of keeping that audience in mind. Ask yourself:

- Who's going to read this?
- How much technical background do they have?
- What things are they interested in?

Your readers will probably belong to one or more of the following groups:

- Industry (If so, what type? And at what level—management, clerical, technical, production?)
- Governing board (board of directors? agency control board? county supervisors? legislative commission?)
- Citizens advisory committee (Do some of them have a technical background? If so, how much? How long has each been a member of the committee? How much do they understand?)
- General public (Will they read this piece in the Sunday newspaper? A weekly magazine? A brochure passed out at a shopping center? Is this piece about a condition or issue they may already be aware of, or are you breaking new ground?)

- Military (What branch? What rank? What occupational specialty?)

Look at that list of groups again and notice that, to some extent, the questions are interchangeable. For example, readers in the "general public" may also be in management or have some kind of technical background. The key to success is to know as much as possible about your readers, their likes, dislikes, background, and biases. This will help you "slant" your writing to the target audience. Make no mistake: Slant doesn't mean building in a bias or attempting to mislead the reader—it's simply the way you choose to look at a subject. Is the glass half full or half empty?

Before you start writing, look for an angle that will help you catch and hold the reader's interest. If your readers are mostly parents, for instance, tell how your topic affects children or show its benefits to the family. If your readers are homeowners, try to discuss the subject in terms of owning and maintaining a home. If your readers are engineers, perhaps you can relate your objective to the importance of doing things correctly, precisely, and on schedule. And you'll always get any audience's attention if you can show them how the product or proposal will help them keep their cost of living down and their standard of living up.

Rule #4: **Know your audience.**

Remember too that all readers of the same document don't bring the same level of understanding to the piece. Even people in the same profession may have different backgrounds, education, or technical knowledge. Make sure the broadest possible range of your target readers will understand you.

Before you start to work on a document, take a minute and try to describe the intended audience in writing. For example:

My audience...

- *Will pick up this brochure at a health fair* (that's where we'll distribute the finished product).

- *Is interested in the problem of reducing air pollution* (otherwise, they probably wouldn't have picked up the brochure).

- *Probably has a limited knowledge of causes of air pollution* (therefore, keep the explanations short and clear; try to relate the objective to things they already know about and understand).

- *May have no knowledge of jargon or acronyms used in this field* (use them only when absolutely essential, and then explain what they mean; perhaps include a glossary).

- *Probably has limited time to read this brochure* (it must be tightly written and it should cover only one topic; keep sentences and paragraphs short).

Exercise #4:
Understanding Your Audience

Think of a type of writing you might have to do—for example, a funding proposal, an instruction manual, a progress report, or whatever you're called to work on. Now, write a description, similar to those on the previous page, of the audience you would be writing for. The more thoroughly you can describe this group, the better you will be able to write for them.

_____
_____
_____
_____
_____
_____
_____
_____
_____
_____
_____
_____
_____
_____
_____
_____

Chapter 3

Thinking

You've defined your objective and identified your audience. Now you're ready to write. Or maybe not. Suppose you can't get started. The words just won't come. You may be a victim of the dreaded and much-publicized affliction known as writer's block.

Writer's Problem #5: Writer's Block

Writer's block is like that scary shadow in the corner of the bedroom when you were a kid. It's not nearly as dangerous as it seems, and if you just throw some light on it, it will go away entirely. "Our present fears are less than our horrible imaginings," said a writer named

Shakespeare. (There's no record of whether he ever suffered from writer's block, but he certainly produced a lot.)

Writer's block doesn't mean you're not thinking. We're always thinking—it's what the mind does naturally. Writer's block means you aren't *writing*. You aren't putting anything down. Thus, the way to beat the block is to start writing. Never say, "I can't think of anything to write." Write something—anything. It's better than just sitting there. Once you start the physical activity of writing, your brain gradually focuses on the topic and begins to supply ideas.

Rule #5: Don't worry about getting something right; get something written.

A good technique is to begin writing words related to your topic. Don't try to organize or prioritize—just put down words as fast as you can. Let your mind wander over the whole subject:

> *...fonts, typeface, soft fonts, scaleable, select, initial fonts, change fonts, print, printer, printing...*

While you're grabbing words out of the ether and slapping them down on the page, a portion of your subconscious will be examining them. Your brain has plenty of capacity to work on two related lines of thought at the same time. Soon you'll find your brain giving you not just words, but ideas. Then the ideas will begin to organize themselves into phrases:

> *...printer, printing, printer memory, need more of it, conserve memory, how? isolate soft fonts, how? load only as needed, how? mark them. mark how?...when printer memory space limited, select fonts that will only be loaded when you need them...do this by marking the name of the font with an asterisk...First, go into "Print Function," then select the font you would like to...*

And so it goes; you're on your way to writing. In fact, you'll find that your brain may be supplying several ideas at the same time, all based on different words you've been writing. Now your problem changes from "I can't think of anything to write" to "Omigosh, how can I get all these ideas down before I forget them?!" Writer's block often occurs because you really aren't ready to start writing: you haven't done your thinking homework yet.

Writer's block can also occur when you try to juggle two conflicting activities—writing and editing—at the same time. If you tried to drive your car in two gears at once, you wouldn't go very far (although your local transmission shop would love you). Writing is another activity you shouldn't attempt in two gears at the same time. Don't try to edit at the same time you write; you'll only worsen your block. Pretty soon you'll be grinding your mental gears and agonizing over every other sentence. You'll quickly bog down in editorial decision making, while your real objective drifts off into the distance like a helium balloon with a broken string.

Keep the big picture in mind. Keep writing toward that final objective. Of course your piece won't be perfect—that's why it's called a "rough" draft. Just keep plugging away, throwing those ideas onto the page, until you come to the end. *Then* you can go back and start fine-tuning the piece. As the Queen told Alice in Wonderland: "Start at the beginning and go until you reach the end. Then stop."

Rule #6: Don't try to write and edit at the same time. That is the stuff of which migraines are made.

After you've captured enough thought-bits and nursed them into full-blown ideas, there are two questions you should ask:

1. **What is the best format to convey this information?** Sometimes the format is dictated by the organization you're writing for; it may require that certain documents be done in certain ways. Other times you'll be able to decide for yourself. In that case, ask yourself what format will get the main idea across quickly, clearly, and effectively: memo? letter? procedure? rules? report? proposal? What format will best help you achieve your objective?

2. **How can I organize the information to best serve my reader?** There are many ways to organize information. Often the material itself will suggest the most logical method for organizing it. You may even find more than one way to do it. Think creatively; sometimes the obvious format is not the most effective. Here are the most common organizing methods:

- **Chronological:**

    In September 1995, we adopted the new policies. Shortly after the first of December, we found that...

- **Step by step:**

    First, be sure the AC cord is unplugged. Next, remove the five screws from the back. Now carefully slide the case off the computer.

- **General to specific:**

    Plants

    Trees

    Pine trees

    Torrey pines

    Torrey pines of San Diego, California

- **Order of importance:**

    Levels of air pollution caused by vehicle engines:

    gasoline (significant)

    compressed natural gas (slight)

    electricity (none)

- **Grouped by similarity:**

    All processes that produce nitrogen oxides

    All household products that contain lead

    All reptiles native to north central Texas

Thinking

Writer's Problem #6:
Organization Not Logical

When most people talk about organizing information, they think of the outline format they were taught in school. There are other ways you can use to organize information to help you in writing or as formats for presenting the information to your reader.

Flowcharts

A flowchart provides a common way to write procedures. In most flowcharts, a *rectangle* represents an action that must be done, a *circle* means caution or a warning of some kind (though sometimes caution is indicated by a rectangle with pointed ends), a *diamond* indicates a decision to be made, a *triangle* flags a cross-reference, and an *oval* supplies related information (in some flowcharting systems an oval means start or stop).

The illustration on the next page shows how a flowchart can help explain a procedure.

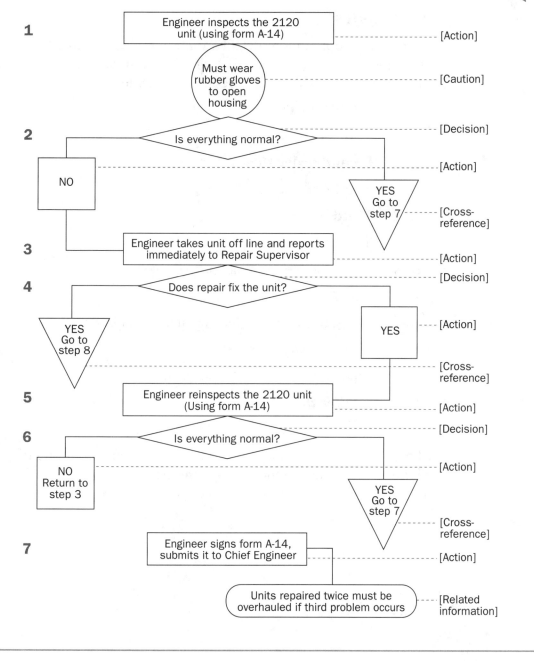

Thinking

Sideheads

Sideheads support a more detailed form of outline writing. Be careful; it's easy to get confused over what is subordinate to what. Notice also which lines are in all capital letters.

1-1. **FIRST-ORDER HEAD**

    Text begins here.

1-1.1. **Second-order Head**

    Text begins here.

1-1.1.1. **Third-order head.**

    Text begins here.

Here's a more concrete example of how sideheads work:

1-1. **PLANT LIFE**

    The second form of life is plants. Botanists often …

1-1.1 ***Trees, Bushes, and Shrubs***

    Among the most useful plants are …

1-1.1.1 ***Coniferales.***

    Conifers are a specialized form of …

Often, if you will spend just a little extra time thinking through your data and selecting the best system for organizing it, you will save yourself a lot of time in the writing process.

Exercise #5:
Organizing Your Thoughts

A. Put a number beside each statement to organize the list into a logical sequence. (The answers are in Appendix C.)

____ Bring me the file folder.

____ Unlock the filing cabinet.

____ Open the right-hand drawer of the desk.

____ The antique writing desk is against the wall opposite the desk.

____ The filing cabinet key is the only silver-colored key on the key ring.

____ The filing cabinet is next to the desk.

____ Find the file folder marked "Harris Plant Floor Plan."

____ Find the key ring in the drawer.

____ The office is the second room on the right; go in.

Thinking

B. Now use numbers to organize the following phrases from the most general to the most specific:

____ Carburetor

____ Ford

____ Carburetor idle-adjustment screw

____ Gasoline-powered vehicles

____ Transportation

____ American cars

____ Grandma's 1967 Ford T-bird

____ Vehicles with wheels

____ T-bird engine parts

Visual Aids

Although they aren't writing in the strict sense of the word, visual aids (graphs, charts, tables, photos, diagrams, illustrations, and even cartoons) are important tools in the technical writer's kit. Visuals should not be an afterthought. When you organize your information prior to writing, that's the time to plan what kind of visuals will help you and where they should go. If well designed, such aids offer an excellent means of transmitting information quickly and in a minimum of space. A good graphic artist is a valuable colleague for the technical writer to have.

Visual devices should be used thoughtfully and never just for show. Don't use visuals for minor, unimportant points or to duplicate information that's clear and understandable from the text. Every such aid should illustrate and clarify some point the reader may have trouble visualizing. Good visuals help pull the reader through the document and make it easier to understand key ideas and their relationships. Here again, it's good to remember the billboard principle: Keep visual aids simple, clear, and bold so your reader can understand the information quickly and easily.

Normally artwork is created larger than it will appear in the final publication; so, be sure it will still be easy to read and understand even when reduced by as much as one-third. It's a good idea to view the artwork through a "reducing glass," which looks like a magnifying glass but actually makes things appear smaller. You can buy one at an art supply store. If you can't read the artwork through a reducing glass, you probably need to have the piece redrawn.

Placement is also important. A visual aid should be placed as close as possible to the concept it illustrates; otherwise, it can be confusing, if not downright irritating, to the reader. All visual aids must be proofread very carefully. Because the typefaces are often small, errors can easily slip by unnoticed.

Line graphs are good for showing variables, but don't try to present too much data in one chart. A busy, overly complex design will only destroy the effectiveness of the illustration. The graph below uses very similar symbols to express very closely related data. The result is almost impossible to interpret satisfactorily, especially when the illustration is reduced for publication:

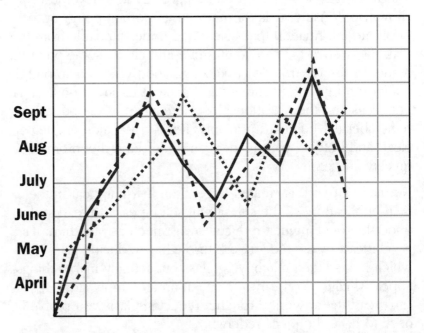

A *bar graph* such as the one that follows would have worked better:

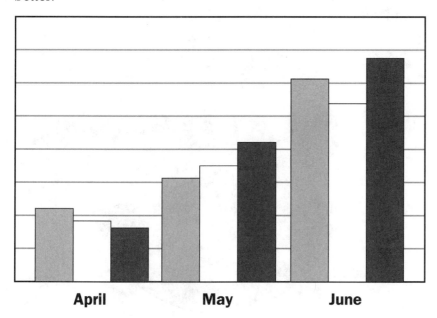

Bar graphs can be deceiving; sometimes bars can mislead the reader into thinking the relationships between items are greater or smaller than they actually are. Be sure the bars are the same width, so the reader will compare data only by the length, not by total area. Bar graphs are best for information that can be expressed in well-defined groupings, such as days of the week or number of widgets manufactured.

When you need to show the relationship of parts to the whole, use a *pie chart* or some variation of it, such as a cube or cylinder divided into sections.

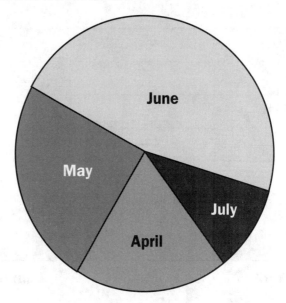

As a general rule, visual aids always have an identifying number, as well as a brief caption that links the visual to the corresponding information in the text. However, visual aids don't necessarily have to be numbered if they aren't referred to in the text. This is often the case with cartoons and with photos or graphics used to divide chapters. It also may be unnecessary to number visual aids, provided:

- Most pages don't have any visuals at all.
- No page has more than one visual.
- The visual is on the same page as the text that refers to it.

When numbering visual aids, you can begin with the first and keep the sequence going until the end of the book, or you can start numbering again within each new chapter. Another method is to give the illustrations sequential numbers that reflect the chapter as well as the illustration. For example, the third visual aid in chapter 8 would be labeled "Figure 8-3."

Sometimes a publication will need different types of visual aids, such as photos, line drawings, charts, and maps. In this case, it's best to number them all sequentially, regardless of type. Here's an example:

Illustration 1. Diagram of air pump components.

Illustration 2. Photo of assembled air pump.

Illustration 3. Graph of pump efficiency at various heights.

If you number them in categories (in other words, Photo 1, Photo 2, Graph 1, Graph 2, etc.), the reader may confuse Photo 3 with Drawing 3.

One more tip: Be sure to use the same terminology in the visuals that you use in the text. If you call something a "filter" in one place, don't call it a "particle trap" in another.

Learning From Other Writers

Some of the best organized and most effective technical writing you'll ever see is done by the authors of children's science books— not the overwritten textbooks you usually find in schools, but the slim volumes in the children's section of the public library. The authors of these books know that if kids don't understand *and* enjoy the books, parents and librarians won't buy them, and publishers won't write royalty checks based on their sales. By the way, children's books are also an excellent way to get a quick overview of a topic before you start studying it in serious detail.

At least once a year, spend some time talking about something technical with a bright nine-year-old. It's a great way to develop the knack of explaining things clearly and in an organized sequence. Kids usually won't let you get away with the kind of gobbledygook and weasel words writers often inflict on readers.

Rule #7: If you write it so a bright nine-year-old can understand it, there's a good chance your boss may be able to understand it too.

Exercise #6:
Learning From the Writing of Others

Go to the public library and spend some time reading children's science books. Analyze the writing and the graphics and ask yourself how you can make your technical writing so clear that the reader can't possibly misunderstand you. Use the space below to record the books you find most valuable, and refer to them often.

_____

_____

_____

_____

_____

_____

_____

_____

_____

_____

_____

_____

_____

_____

_____

_____

Chapter 4

Writing

Once, during a master class with the great violinist Jascha Heifetz, a student was trying unsuccessfully to explain an idea. Frustrated, he burst out, "No, Maestro, you don't understand."

"Stop!" Heifetz interrupted. "Never say, 'You don't understand.' Instead, you should say, 'Maestro, I failed to make myself clear.'"

As a professional writer, you are never allowed to blame the reader for any lapse in communication. If the reader doesn't understand, it's your fault. The burden is on you to make yourself clear and to get your ideas across.

Keys to Getting Your Idea Across

1. **Keep your reader's needs in mind.** If this advice sounds vaguely like our old friend Rule #1, it's supposed to. The reader's needs should be the most important concept that propels the technical writer.

2. **State the most important idea right at the beginning.** You've probably received a phone call (usually during dinner) from someone who calls you by your first name and says something like, "Hi, this is Chris. How ya doin' tonight?" While you struggle to remember who Chris is and where you met, the caller asks you to "answer a couple of questions to help us with a survey." After you do so, the voice then offers to "let you in on a great opportunity." At this point, you realize this is only a poorly disguised sales pitch, and you're angry that someone trapped you and wasted your time.

   There's a good chance the person who reads your report is also, to some extent, a captive audience, so get to the point, right in the upper left-hand corner of page one. Tell your reader what the main idea is, why it's important, and how it may affect him or her. Remember your experience with the telemarketer, and don't irritate your reader.

3. **Write the first draft the way you talk.** Pretend you're explaining something to a friend, face to face. Don't try to be fancy, don't try to impress anyone—just get the ideas down in clear, understandable language. Naturally you want the vocabulary to be appropriate to the audience, but the first draft isn't the place to worry about that. Write it the way you would say it. Then go back later and polish as necessary to fit your audience. If you're still not convinced, read these two sentences:

   Delineate inscribed confabulation in a form proximate to the fashion used to verbalize the original perception.

   Or—

   Write it the way you say it.

   You decide.

4. **Include all the necessary information.** Early in their training, journalists learn something called "the Five W's of Journalism" (there's also an H, but "five W's and one H" just doesn't have the same ring). They are *who, what, when, where, why,* and *how.* These are the things the reader needs to know, and leaving one or more out almost guarantees misunderstanding. When a subject is unusually complex you may want to make a list of all the points that need discussion so that as you write you can check each off to confirm you've covered them all.

5. **Make sure the information is understandable.** Have you ever had to go back and read the same paragraph two or three times to understand what it means? You probably resented having to take the extra time, and your readers will resent it too. Never make a reader go back and reread; make the message clear the first time.

6. **Use only one main idea in each sentence.** You want modifiers, adjectives, and dependent clauses? Fine—just make sure they all refer to the same idea. You should also be careful not to connect independent clauses with a comma, that's called a "comma splice," this sentence has two of them.

7. **Keep your sentences short.** Technical writing often demands that you talk about new technology, concepts, and methods, so it's quite important that every sentence be crystal clear. Remember that shorter sentences are easier to understand. Good technical writing often produces sentences with only eight, ten, twelve, or fourteen words. If you really need twenty-two words to get the meaning across, use them, but be very careful to make every word count. And don't write long, complex sentences that have several clauses. Keep the structure simple.

8. **Keep your paragraphs short.** You may remember hearing this rule somewhere: When you start a new idea, start a new paragraph. True, but you can also start a new paragraph on the same idea if a paragraph is becoming too long. Studies have shown that people can read the same material faster and understand it better when it's presented in short paragraphs. In general, no paragraph should exceed about ten lines. You can always find some place to break a long paragraph into two short ones. In fact, the longer it is, the easier it will be to find a breaking point.

Perhaps your high-school English teacher also taught you another rule: Never write a paragraph that has only one sentence. Generally, this is a very good rule, but it isn't inflexible. If you're careful not to do it very often and if you do it only when it isolates an idea, writing a one-sentence paragraph can be an effective way to focus the reader's attention.

Anything that helps you communicate with the reader is worth considering.

Exercise #7:
Testing the Clarity of Your Writing

Suppose you could give your readers a test on something you wrote. Would they be able to explain your main point in three or four clear sentences? Use this exercise as an opportunity to test your own writing abilities. Choose a memo or something else you've written recently. For simplicity, select something that's one page or less. First read a sentence or paragraph to someone who isn't already familiar with the content. Then ask your listener to tell you what it meant. If the result is a reasonable facsimile of what you intended to communicate, give yourself an A+. If it's not, use the lines below to rewrite the problem sentence or paragraph.

_____

_____

_____

_____

_____

_____

_____

_____

_____

_____

Hooking the Reader

Have you been quickly drawn into an article or proposal recently? The "hook" is the opening of a piece, usually the first sentence, or perhaps the first two sentences. A good hook makes you want to keep reading. Here are a dozen ways to create an interesting opening:

1. **Summary.** Condense your main idea into one sentence (perhaps two). The rest of the piece will then provide explanation, detail, and supporting information.

   Effective meetings are a lot like trains. They should leave on time, follow an established route, and deliver passengers safe and sound to the destination they're expecting.

2. **Grabber.** Use a statement, detail, or statistic that grabs the reader's attention:

   Oil-based paints contain up to five times more toxic ingredients than water-based paints.

3. **Quotation.** An eloquent or interesting turn of phrase from a well-known figure can offer a thought-provoking opener:

   Albert Einstein once wrote, "God does not play dice with the cosmos."

4. **Question.** You can ask a question that summarizes the topic:

   Are personal computers making us fat?

   Or pose one that is answered later in the piece:

   Can you name three sources of deadly fumes that may be in your office right now?

5. **Opinion.** Associating an idea with an authoritative or controversial person can give it instant "personality":

   Senator Robert Hicks believes a new law will force hundreds of non-polluting factories to close.

6. **Prediction.** This sort of opener tantalizes the reader with a peek into your crystal ball:

   Analysts say that by the year 2000, our industry will be in the middle of a severe downturn.

7. **Description.** Few readers can resist reading on to discover the answer to an opening "riddle":

   It's blue, black, or gray; turns the sky brown; and makes the summers hotter.

8. **Anecdote or story.** You can inject a sense of realism or relevance with a true story:

   A student was once asked to stop coming to school because he just couldn't make himself conform to the school's policies and practices. The boy's name was Albert Einstein.

9. **Analogy.** An easy-to-grasp analogy can tame big ideas while at the same time offer a potent image:

   A neutrino is like a B-B in a universe full of beach balls.

10. **Definition.** Defining your key concept at the outset slides your reader quickly over the first hurdle:

    Post-reduction: The reduction of chromosomes in the second meiotic division.

11. **List.** The advantage of a list is that it immediately both highlights and defines the important features of your piece:

   There are six qualities every chemist needs: patience, perseverance, caution, perseverance, imagination, and perseverance.

12. **Humor.** Be careful: Anytime you try to embellish a technical piece with humor, have at least one other person read it. Make sure it really is funny, that it is appropriate in context, and above all, that it won't offend anyone.

Exercise #8:
Writing an Effective Opening

Practice writing an interesting opening sentence. Use your own topic or pick one of these:

- A brochure informing the public about an 800 number they can use to report vehicles that are polluting the air

- An article about the need for increasing the amount of money for establishing and maintaining sanitary landfills

- A report to stockholders explaining that there will be no dividends this year because the company needs to modernize the St. Louis factory

- A piece explaining to technicians the importance of maintaining equipment on a regular basis

_____

_____

_____

_____

_____

_____

_____

_____

_____

_____

Getting Out Gracefully

You may have been taught somewhere that a report has three parts: introduction, body, and closing. That's a good way to organize a document, but it isn't the only way. Sometimes your introduction may only be the first paragraph, or perhaps just the first two or three sentences. It may flow so smoothly and quickly into the body of the document that the reader doesn't think of it as an introduction.

The same is true of the ending, often called the "exit" or "close." It may be a formal summary, complete with a side heading that says "Summary." On the other hand, it may be only one or two sentences that seem to form as the logical result of everything that went before. As a general (but not inflexible) rule, the longer the document is, the longer your closing will need to be. Supposedly someone asked Beethoven why his fifth symphony ended with the same crashing chord repeated seven times. He replied, "Because that's how long it takes to resolve the musical tension that has been building up throughout the symphony." If your document is lengthy and has covered many issues, the closing may require several paragraphs to ensure that the reader remembers all the points and understands their relationship.

Even in long documents, be careful not to write more than is necessary. Ideally, your close will be short and to the point. Sometimes it helps to think of a document as having the same structure as a joke. A good joke consists of two parts: the set-up and the punch line. If the set-up is too long, the listener will get bored and the punch line will lose much of its effect. If the punch line is too long, it loses steam and the joke falls flat. Many documents can reflect this same two-part structure. That is, the first part offers a brief introduction that flows into the main body, and the second part (the closing) caps the piece with impact and "punch."

However you write it, the close should tie everything together and bring the reader back to the original starting place, which is the objective you set for the document.

Transitions

It's important to write interesting hooks and closes, but they won't save a piece that's poorly written in the middle. One of the most important devices for holding a reader's attention throughout the piece is the *transition*—the connective tissue between ideas and paragraphs. These words and phrases link paragraphs together and smooth ideas into one seamless whole. Transitions help your writing move from one thought to the next in a logical sequence. Following are some ways to accomplish this:

- End a paragraph with a question, or with a statement that leaves a question in the reader's mind. The reader will keep going in order to learn the answer.

    ...agree that we will always need meetings, although many managers ask, "Why can't meeting time be used more effectively?"

    The secret of running an effective meeting, according to a study by the Harvard School of Business, is based on the principle of...

- Start a new paragraph by repeating a key idea from the preceding paragraph.

    ...the potential danger of letting technicians make adjustments to such an expensive piece of equipment.

    When technicians are allowed to "personalize" the unit to fit their individual working characteristics, it doesn't have to spell disaster.

- Repeat a key word from the preceding paragraph.

    ... "Sometimes we just can't tell what the designers intended," one employee said.

    What designers intend, in many cases, is for the personnel in the field to...

- Use an idea in one paragraph as a launching pad to propel you toward a different idea.

    ...thanks to new technology that makes the unit much more portable than the previous model.

    Portability is not the only thing to be considered, however. For some users, the unit's small size can actually be a drawback...

- Begin the next paragraph with a connective word or phrase.

    Another point to consider is...

    As a result, ...

    At the same time...

    Consequently, ...

    First (second, third, etc.), ...

And finally, …

For example, (or, for instance,)…

For this reason, …

However, …

In fact, …

In the same way, …

Naturally, …

Next, …

Obviously, …

On the other hand, …

Yet…

Naturally, you shouldn't use the same kind of transition too often in one piece. Obviously, it would call attention to itself. On the other hand, transitions are important to achieving a smooth flow. Yet in the same way, however, for this reason, a good writer should…(uh-oh, perhaps it's time to move on to another topic).

Sentence Length

Although it's good to keep sentences short, too many brief bursts can make your writing seem choppy. Vary the sentence length to hold the reader's interest. Short sentences can often be connected with punctuation or connective words, but *only* if the sentences discuss the same idea (or very closely related ones).

Choppy:

Carbon dioxide is a "greenhouse gas." It traps radiation in the atmosphere. It prevents its passage into space. This results in a warming of global temperatures.

Better:

Carbon dioxide, a so-called "greenhouse gas," traps radiation in the atmosphere and keeps it from passing into space. Trapped radiation is a primary cause of global warming.

Overly-complex, too many ideas:

Carbon dioxide is one of various "greenhouse gases" known to trap radiation in the atmosphere, thereby preventing the radiation from passing into space, resulting in a general warming of global temperatures, and representing about half the nation's total cost of waste disposal.

Rule #8: A sentence should have only one main idea.

Gender-Neutral Pronouns

Language is a living thing, and as such it changes over time. Words come, words go. New ways develop to handle new situations. Usage once considered acceptable turns quaint and outmoded, or perhaps is abandoned entirely. One of a writer's responsibilities is to keep up with changes.

Among the more recent changes in English is the growing sensitivity to gender-biased pronouns. Writers in response are becoming more attuned to the fact that not every person referred to is a *he*.

This presents a challenge. Traditionally, when referring to a person in a male-dominated profession, writers wrote *he*. When discussing someone in a female-dominated profession, writers used *she*. But now writing must reflect the recognition that not all doctors are male, not all nurses are female, the "fireman" who saves someone's life may be a woman, and the "Secretary of the Year" could easily be a man.

Some writers start off on the wrong foot by apologizing in the introduction: "In this volume, we will use the traditional pronouns *he, his,* and *him,* although we hasten to assure the reader that we are indeed sensitive to…" and so on. It's a cop-out.

The obvious solution would be to invent a new pronoun that would be generic and gender-free. But what? *Hesh? Herm?* Many have tried, but alas, no one has yet devised a neutral pronoun that doesn't sound awkward or even downright silly. Until someone does, there are other ways the thoughtful writer can use to keep "hermself" from committing a blunder.

One way is simply to write *he or she, his or hers, him or her.* Unfortunately, this device will work only for occasional usage. If you use this combination more than once or twice on the same page, it calls attention to itself and detracts from your message.

Some writers insert the diagonal or slash mark instead of using the word *or* (for example, *he/she* or *his/her*), but many readers won't understand what this means. The diagonal has two meanings. It can show alternatives ("It's an either/or situation"), but it can also be used to combine functions ("Penny Marshall is an actor/director"). For this reason, it's probably safer not to use the diagonal.

When we talk to each other, many of us unconsciously handle the generic pronoun problem by using the words *they* and *them*. This is also acceptable in writing, provided you're careful that the subject agrees in number with the pronoun (*they* and *them* are plural, so make sure the subject of the sentence is too). Here's how the creation of a gender-neutral sentence might progress:

> *The owner of a small business often wonders why his profits seem to vanish by the time he gets his overhead paid.*

This isn't sensitive to gender; rewrite it:

> *The owner of a small business often wonders why his or her profits seem to vanish by the time he or she gets his or her overhead paid.*

Now it's gender-neutral, but awkward. Rewrite it:

> *The owner of a small business often wonders why their profits seem to vanish by the time they get their overhead paid.*

Oops! The subject and the pronouns don't agree in number. Make the subject plural:

> *Owners of small businesses often wonder why their profits seem to vanish by the time they get their overhead paid.*

At last you have a sentence that is both gender-neutral and grammatically correct. Another option is to rewrite the sentence using the second person pronoun you:

If you own a small business, you probably wonder why your profits have vanished by the time your overhead is paid.

Or rewrite it to eliminate the gender problem entirely.

The overhead is paid, but there are no profits left—a common problem for owners of small businesses.

You can always rewrite something, and when you do, it nearly always improves.

Exercise #9:
Practicing Gender-Neutral Writing

Rewrite the following sentence in at least three ways, all gender-neutral:

Whenever a technician repairs a unit in the field, he must submit Form 72/30 to the Accounting Department.

1. _____

2. _____

3. _____

Writing About People With Disabilities

You will be less likely to offend if you use the term *person with a disability* rather than *handicapped*. This puts the person ahead of the disability and helps to maintain the sense of dignity that every individual is entitled to. The term *disabled people* is less desirable, but you may occasionally have to use it in order to avoid awkward sentence structure. Try not to use the term *handicap* when referring to people. Most people with disabilities prefer to use *handicap* to mean the environmental conditions, such as narrow halls and steep ramps, that make life difficult for them.

You may find that your organization has a style rule already in place that deals with this writing challenge, and you will simply have to do it that way, even though it may be unwieldy.

Chapter 5

Editing

When you edit, you're not looking for typos, misspellings, and wayward semicolons; that's proofreading. Editing is rewriting, polishing, tightening, improving. Editing is asking yourself: "Will this piece accomplish the objective? Can I make it work better?"

Never deceive yourself into thinking your first draft is "good enough." Everything needs to be edited, even memos and short notes. In fact, here's a piece committed to perpetuity (an epitaph) that would have benefited from some editing:

> "Sacred to the memory of Major James Brush,
> killed by the discharge of a pistol
> by his orderly, 14 April 1831.
> Well done, good and faithful servant."

Rule #9: **Never send out a first draft. Rewrite everything!**

When you've finished the first draft, put it aside for a day or two, if possible. Give yourself a chance to forget it. When you come back and read it, you'll see it more like your reader does, you'll be more objective, and the weaknesses will be easier to spot.

After you've let the piece rest awhile and you're ready to rewrite, don't immediately start changing things. First read it through from top to bottom without stopping. As you do, ask these questions:

- Is the message clear? Does the objective come through clearly?

- Does it have a good "hook"? Does the opening grab the reader's attention?

- Does the piece get to the point right at the beginning? Will the reader understand immediately what it's about?

- Is the style of writing appropriate for the audience? Is it too technical? Or not technical enough?

- Is the vocabulary appropriate for the reader? (See Appendix B to learn how to use the Gunning Fog Index.)

- Does it have the right tone?

- Does the body of the piece develop the main idea? Did I avoid going off on tangents? Does the piece stay on track until the end?

- Is it tight? Did I eliminate unnecessary words, phrases, sentences? Did I use language economically?

- Does it flow? Does each idea lead to the next in a logical sequence? Do the transitions pull the reader along?
- Does the ending tie everything together?

Be Aggressive About Passive Voice

One of the most common writing weaknesses to watch for when you're editing is the use of passive rather than active voice. In the active voice, the subject of the sentence does the acting:

The committee made a poor decision that could hurt the company.

Here the "actor" was the committee.

In the passive voice, the subject is acted upon:

A poor decision was made that could hurt the company.

In this example, the reader doesn't know who or what made the decision (and that lets the committee off the hook).

Passive voice is weak because it doesn't tell the reader who took the action. It's a roundabout way of expressing an idea. You can usually spot passive voice by looking for the presence of, or at least the implication of, the idea "by someone":

This report was written to explain why organizational right-sizing is deemed necessary.

So who did what to whom? The report was written [by me]. The action was deemed necessary [by the management]. Translation: "I wrote this report to explain why the management has decided to lay people off."

There are, however, some occasions when it is acceptable to use passive voice, particularly in technical writing:

- When the actor isn't important to the topic and might even detract from the focus of the document.

    The sample was washed and prepared for mounting.

    NOT

    Lab assistant Terry Smith washed the sample and prepared it for mounting.

    Most writers wouldn't want the reader to be distracted and waste time wondering who Terry Smith is. On the other hand, the name should be used if the person is the subject of the document.

    Dr. George Washington Carver examined the results of his experiment carefully.

    In procedures or instructions, you can avoid passive voice simply by using imperative sentences.

    Clean the sample thoroughly with denatured alcohol; then heat it in the autoclave for three minutes.

    NOT

    The sample is cleaned with denatured alcohol; then heated in the autoclave for three minutes.

- When naming the actor would not be diplomatic.

    An honest mistake was made, but we have taken steps to ensure that it can never happen again.

    In this case, there's nothing to be gained by pinning the blame, so the writer decided it would be kinder (and probably smarter) not to identify those who made the error.

- When it isn't possible to name the actor.

    The window was broken sometime over the weekend.

    Here the writer would be happy to pin the blame, but the culprit is unknown.

- When the subject of the sentence is more important than the actor.

    Their phone number was listed in the yellow pages.

    Putting that sentence into active voice would result in a statement that gives the telephone directory an unwarranted significance:

    The yellow pages listed their phone number.

    Here's another example:

    The President was detained by a flat tire.

    Presumably, the President is more important than a tire, flat or otherwise.

Rule #10: The key to avoiding passive voice: Write about someone doing something (and in that order).

Exercise #10:
Stamping Out Passive Voice

Rewrite this sample to eliminate the passive voice:

Some pollution is caused by a condition known as temperature inversion. Because the city is surrounded by mountains, cold air is trapped near the ground by warm air as it comes over the mountains. Then heat from the ground is radiated back into the atmosphere and the cool air is forced up. The polluted air is still trapped inside the valley, however, and pollution builds until it reaches unhealthful levels.

Rethinking

Good rewriting often requires rethinking. If you keep thinking about something in the same way, you'll probably keep writing about it in the same way. It's a good idea to ask yourself how the idea would look if you approached it from a different direction: upside down, reversed, inside out. Don't be afraid to stand an idea on its head. Let's take another look at the paragraph you worked on in the previous exercise.

> *Some pollution is caused by a condition known as temperature inversion. Because the city is surrounded by mountains, cold air is trapped near the ground by warm air as it comes over the mountains. Then heat from the ground is radiated back into the atmosphere and the cool air is forced up. The polluted air is still trapped inside the valley, however, and pollution builds until it reaches unhealthful levels.*

Imagine that this sample comes from a brochure on the causes of air pollution. The purpose is to educate readers and motivate them to reduce air pollution. The audience is any average adult, perhaps someone who picked up the brochure at a county fair, shopping mall, or similar place where organizations have information booths. Now, rethink the main points about temperature inversions with the help of the diagram on the next page.

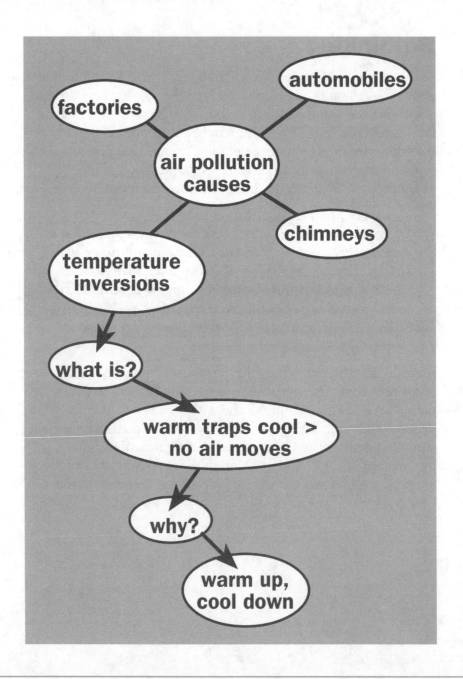

If you look at the idea in the usual way, it's very logical to start with *temperature inversions* and then work in the direction of the arrows. This in turn pretty much forces you to begin your paragraph by explaining what a temperature inversion is. (Because you're writing for a lay audience, you can't assume they'll know. Rule #1, remember?). Maybe this explains why so many brochures and articles begin with a definition of a term: The author hasn't tried to look at the idea from various angles.

Rule #11: Don't be afraid to look at the topic from different angles.

Let's approach the idea from a different direction. Instead of starting at the beginning, look at the final thought—*warm air rises, cool air sinks*. Remember that a key to good teaching is to start with what your audience already knows. While most people don't know exactly what a temperature inversion is, they *have* heard somewhere that warm air rises. It's the reason steam from a kettle, cinders from a fire, and smoke from a skillet all go up. Readers can visualize those examples. It's something they understand.

Suppose you use the end as your starting point and put the idea into an example most people can easily visualize: Warm air is rising from a container; then someone puts on the lid. Using this line of logic, you can write an entirely different paragraph:

> *Warm air rises and cool air sinks. At night, the valley is like a jar filled with polluted cool air. When warm air comes over the mountains the next day, it acts like a lid, trapping the cool air below. This is a temperature inversion. Later, warm air rising from the ground will push the polluted cool air upward a little, but it will stay trapped inside the valley and become smog.*

Some readers would be comfortable with the first version of this paragraph. But many others would find the second version easier to understand. It's your job to know your audience and make sure your writing works for *them*.

Rule #12: People learn better if you start with what they already understand. Move from the known to the unknown.

Exercise #11:
Rethinking Your Approach

Rethink the two temperature inversion paragraphs. Try to find another way of approaching the topic, one that might make the message clearer. Write your version here.

_____
_____
_____
_____
_____
_____
_____
_____
_____
_____
_____
_____
_____
_____
_____
_____
_____
_____
_____
_____

Editing

Eschewing Obfuscatory Utterances

Technical editors sometimes use the expression *COIK,* which means "clear only if known." In other words, the reader will understand your writing only if he or she is already familiar with the subject or the terminology. This is dangerous territory. Don't just assume your reader will understand: Make sure of it! Write with the least technical member of your audience in mind. When in doubt, write for your grandmother! (Unless she is a noted nuclear physicist.) No one ever complains, "I can't understand this document; it's just too clear and concise."

Jargon

Jargon is language specific to a certain field or profession and often makes communication easier or clearer among members of that field. When the reader comes from outside the field, however, jargon just creates fog. Jargon undermines the primary function of good writing, which is to explain something in terms the reader will understand.

Job-related slang is another form of jargon:

> *I picked up a pinch-book and got on my motor to go out on 15 and balance DUIs.*

Translation:

> *I picked up a traffic citation book and got on my motorcycle to go out on Highway 15 and administer the field sobriety test to drivers who appeared to be driving under the influence.*

If you haven't already realized it, that example was supplied by a state highway patrol officer.

Technical and Scientific Terms

If you are positive your reader will understand a term, use it. But if you have the slightest doubt, either use a more common term or explain what the technical term means. Don't assume someone will know what a Magedeberg hemisphere is just because you do. Even among scientists and technicians, not every reader has the same background.

Trendy Buzzwords

A certain executive loves to begin letters with "Re our telcon..." He thinks it's a really nifty shorthand way of saying "Regarding our recent telephone conversation." True, "re our telcon" is shorter, but it has one rather serious flaw: Most people won't understand it. They may figure it out on the second or third reading, but forcing your reader to decipher cryptic phrases is not good writing. Some recent examples of buzzwords include *reality check, perception check,* and *paradigm shift.*

Here's a terrific (and terrible) example of succumbing to the temptation of buzzwords:

> *As the Generation X experientially impacts the Information Society, we may see a paradigm shift in our conceptualization of parenting within the nuclear family.*

Borrowed Language

Don't adopt or adapt terms and expressions from other fields unless there is a clear advantage in doing so. Avoid using expressions that aren't yet fully accepted into the language and widely understood. It will only detract from your message and probably offend readers who value plain talk.

Editing

Examples creeping into common use from computer science, for example, include *to access, to interface,* and *downtime.*

Not   *The guard can access the lab with his keys.*

But   *The guard can get into the lab with his keys.*

Or   *The guard can gain access to the lab with his keys.*

Verb-ification

Resist the urge to turn nouns into verbs, as in "How is this going to impact the committee's work?" We didn't need anyone to change the noun impact into a new verb, *to impact*. We already have a perfectly good verb for this purpose, *to affect*, which means the same thing and is more widely understood.

Here are some other examples:

- To *conference*

    The teacher said he would be conferencing with parents individually. (Use instead *he would confer* or *he would consult*.)

- To *transition*

    The industry is transitioning into a new era. (Just say *The industry is moving...*)

Acronyms and Abbreviations

An acronym is an abbreviation that can be pronounced like a word. It may be taken from the initial letters of words, such as MADD (Mothers Against Drunk Driving), or it may be made up of parts of words, as in the case of Nabisco (the National Biscuit Company).

When using acronyms and abbreviations, follow the "rule of first citation": The first time you use an abbreviation in your document, spell the expression out and put the abbreviation in parentheses immediately following; after that, you may use only the abbreviation in the rest of the document.

> *The Board of Veteran Appeals (BVA) is reviewing the case. In this instance, the BVA may need four weeks to issue a decision.*

Sometimes the sentence will read more smoothly if you use the abbreviation first, and then the full term. This is often the case when the abbreviation is more widely understood than the term.

> *Samples of DNA (deoxyribonucleic acid) can be taken from a variety of sources. DNA exists in the cells of all living things.*

Sometimes a lengthy report will contain several sections or chapters, and certain readers will need to read only certain sections. If this is likely, then you should identify the abbreviation the first time you use it in each section.

Finally, you may omit the "rule of first citation" in cases where the abbreviation is so well known that there's no likelihood your reader will misunderstand *(FBI, USA, NATO)*, or where the original meaning of the abbreviation has generally been forgotten *(radar, laser, AFL-CIO)*.

Chapter 6

Proofreading

It's your responsibility to proofread your own work, and that means everything. Don't overlook the small print in graphs, charts, illustrations, tables, or captions. No matter how small the error, someone will spot it; that's guaranteed.

Keys to Accurate Proofreading

The secret of good proofreading is accuracy. Here are a half-dozen tips to help you:

1. **Use a straightedge.** Lay a ruler, folded piece of paper, or index card under the line you're proofreading to force your eyes to stay on that line until you finish.

2. **Use the "buddy system" if possible.** Take turns reading out loud with another person who also has a copy of the manuscript. You keep each other alert that way and spot more errors.

3. **If you write on a computer, proofread from a printout as well as from the screen.** Writers often miss typos on the computer screen that jump out from a printout.

4. **Keep your mind alert by playing tricks on it.** Make an enlarged photocopy of the document and proofread that. Because it looks different from the original, your brain will be more alert. Another trick is to proofread in a different place from where you write. Sit on the opposite side of your desk to do your proofreading, or move to a different spot in the room (or even a different room).

5. **Cover the entire page with a blank piece of paper so you can see only the last line.** Start with the last word and proofread one word at a time going backward up the page. You won't catch mistakes that have to do with meaning, but you also won't be distracted by the meaning. You often can spot spelling and capitalization mistakes this way.

6. **Proofread for no more than about fifteen minutes at a time.** Your brain gets tired of concentrating during marathon proofreading, and you lose accuracy. If possible, take a short break three or four times an hour (at a natural break in the manuscript). Perform some other activity for a few minutes, such as opening your mail. If it isn't practical to take a real break, at least take your eyes off the paper and think about something else for a minute. This will rest your eyes, break the monotony, and help you stay effective.

Proofreading Technical Terms

Many technical, medical, and scientific terms come from the names of scientists, such as *gauss meter* (Karl Friedrich Gauss) and *brownian motion* (Robert Brown). However, although the surnames themselves are capitalized, the scientific terms based on them may or may not be capitalized, as in the following example:

> *hertz oscillator,* but *Herzian wave*
>
> *parkinsonism,* but *Parkinson's disease*

To further confuse matters, the abbreviation for *megahertz* capitalizes the first letter (MHz), but kilohertz doesn't (kHz). If you're not sure about the spelling of a technical term, consult a dictionary or reference book specific to that field. The same holds true for names of biological species, geographical sites, chemical compounds, and scientific equipment. There is bound to be a reader who will notice that your *MacDougall furnace* should have been spelled *McDougall*.

Proofreading Marks

When you're proofreading on the computer, you can simply correct mistakes as you go, but when you correct a printout that you (or someone else) will later retype, there's a little more work involved. Here you will need to use proofreading marks.

Be sure that your marks in the text are small, neat, and clear. Don't obscure other parts of the text by haphazard marking. This is where it really helps to have a steady hand and a sharp pencil or fine-tipped pen. Be sure to use a color other than black so that your markings stand out clearly.

Proofreading

In general, for every proofreading mark in the body of the text, a corresponding mark should appear in the margin. If it isn't possible to write something neatly and clearly in the text (added words, for example), just use the caret mark (^) at the spot where you want the correction, then explain in the margin what you want, or use the proofreading mark. The marginal mark alerts the typist to look for the problem in the text and explains what you want added or changed. The caret must show exactly where to put the change. (The word *caret* is Latin for "it is lacking.")

The following listing shows the proofreading marks you will need most often. There are others, but these will cover almost all the situations you're likely to meet. For a more complete list, check a source such as *The Gregg Reference Manual*. While the proofreading marks shown here are fairly standard, the others vary somewhat and all reference books don't necessarily agree on them.

In the Margin	In the Text	Means
↱	He said, "I'm late?"	insert a comma
↻	He said, "I'm late?"	insert an apostrophe
↯ or ↻↻	He said, "I'm late?"	insert quotation marks
? ∧	He said, "I'm late"	insert a question mark
⊙	He said, "I'm late"	insert a period
# ∧	He said, "I'm late."	insert a space
2# ∧	He said: "I'm late."	insert two spaces
∧ or ⋏	He sad: "I'm late."	insert a letter
∧	He said: "I'm late."	insert a word
ꝗ or ___	I'm very late.	delete a word
⦸	I'm shure I'm late.	delete a character and close up the space
⌒	I' m sure I' m late.	delete space
⁋	sentence. A new	begin new paragraph here
∧	H2O	subscript
∨	footnote.3	superscript
] [	]title[	center the element (word, title, etc.)
(stet)	He said, "I'm very late."	you deleted something, then decided to leave it in
=	I'm filled with self-pity.	insert a hyphen
( )	(I guess I'm always late.)	insert parentheses

Proofreading

In the Margin	In the Text	Means
[ ]	[Editor's note: Buy a watch!]	insert brackets
l.c.	He said, "I'm /Late."	lowercase a letter
caps	He said, "i'm late."	capitalize a letter
1/m	He said, "I'm late/catch me when I get back."	insert a dash
tr.	He siad, "I'm late."	transpose letters or words
sp	They live at 1234 Post Dr.	spell out

Occasionally, you'll want to add something lengthy and there won't be enough room for it. In that case, write *see note* Ⓐ or just Ⓐ and put the addition at the bottom of the page or on a separate page (identified as Ⓐ or whatever is appropriate).

Grammar and Spelling Software

Don't grow dependent on software to catch your errors. Grammar checking programs are slow to use and in many cases not very helpful. These programs often flag something in the text and then ask you if you've made a certain grammatical error. If you knew that, you wouldn't need the software!

Often such software also will say things that are downright silly. It will locate *Appendix A* and tell you not to end a sentence with the letter *a*. As it happens, *Appendix A* isn't a sentence, it's a heading (and by the way, that last sentence ended with the letter *a* and it was perfectly all right).

Spell checkers are better, but even after using one, you must still check your own spelling. A spell checker only confirms whether a certain spelling is in its dictionary; it can't know whether you used the right word or not. The following poem is free of spelling errors, at least according to the spell checker:

Memo to My Spell Checker

Eye no ewe don't believe me,
butt you make me type miss steaks.
At leased, it seams two me yew dew,
Four I all ways spot some fakes.
You'll pass on *hair* when I mean *hare*
and *bite* when I mien *byte*.
You'll let me air when I write *heir*,
so my spelling's an awful site.
Now, fair is fare, I quiet agree,
but rite should still be right.
Sew perhaps sum day wee to should meat,
and talk, just you and aye.
We'll cheque our fax and tri to beat
this problem, dew or die.
Fore hour English language, as you sea,
makes spelling quite a fright.
Its bin too month's sense last I saw
a word I could spell write.

©1994 Robert McGraw

If you are writing material for an English-speaking country other than the United States, remember that nearly all spell checkers use the U.S. spellings.

Exercise #12:
Improving Your Proofreading Accuracy

How's your proofreading skill? See how many punctuation and spelling errors you can find in these sentences. The correct versions appear in Appendix C.

1. We cant glue it back together but maybe we can still do the test

2. The video project requires four items a letter of agreement a treatment a first draft and the final script

3. Although no one ever heard of him Dr Monty had written many scholarly articles including The Structure of Polymorphonuclear Leukocytes

4. After writing his book Polymorphonuclear Leukocytes, Elvis and the JFK Conspiracy Dr Monty became a well-known guest on TV talk shows

5. Jerry replied What I said was Its well known that Dr Monty wrote Polymorphonuclear Leukocytes Elvis and the JFK Conspiracy. Didn't you understand me

6. The recommendations team David Boric Dr Karen Joyce and C Phil Morris was highly qualified to analyze the Copperfield project

7. The commitees reports vetoed Karens suggestion for these reasons its too costly the time so they said wasnt appropriate and it wasnt there idea so it couldnt be any good

8. Karen quit and started her own company she is now a millionaire Dr. Monty taking a page from Geraldos book got his own talk show where he interviews vampires witches and politicians ex wives

Chapter 7

Some Common Types of Technical Writing

Up to this point, you've learned the principles and techniques that are important to all technical writing. Now you'll look at some hints that apply specifically to the most common types of technical documents.

Instructions

The following suggestions also apply to procedure manuals, which are often a collection of several sets of instructions.

- Don't omit articles *(a, an, the)* or pronouns, as in this example:

    Hold unit in left hand with top toward body and compartment open. With wrench in right hand, tighten nut on black module on right side of compartment.

It doesn't take much more paper to say "With a wrench in your right hand, tighten *the* nut on *the* black module on *the* right side of *the* compartment." But it's easier to understand, especially if the reader has never done this job before.

- Don't backtrack; some readers may attempt to follow the instructions without first having read them entirely.

    Step 4. Remove the access panel.

    Step 5. Remove the retaining screw on the oil reserve tank.

    Step 6. Catch the old oil as it flows out. Because the tank holds more than it appears to, you should have brought a container large enough to catch the oil (one-gallon capacity is sufficient).

    Oops! The reader is now standing there holding a pint jar full of oil while the rest drains onto the floor.

- Use illustrations where appropriate—good ones often greatly reduce the number of instructions you need to write.

- Use the imperative, not the indicative, mood. The indicative states a fact; imperative mood gives a command or makes a request.

    Wrong:

    The wiring harness should be removed from the control module.

    Right:

    Remove the wiring harness from the control module.

- Include a trouble-shooting page that lists most common problems or errors along with their solutions.

Proposals

The main purpose of a proposal is to persuade the reader to hire your company, buy your product, or fund your project. Most result from an RFP (request for proposal), a document distributed by organizations soliciting proposals. Obviously, you should read the RFP carefully before preparing your proposal, making sure you understand what the potential client needs and how you can satisfy that need.

There's another reason to read the request carefully, a reason many writers overlook: to analyze the writing style. By writing your proposal in a similar style, you will send a subtle but powerful message that you can think like the reader and fit in with the reader's organization. Once you understand what the RFP is about, analyze the style. How are the ideas organized and presented? How technical is the vocabulary? How long are the paragraphs? If you can write your proposal in a similar style (*and* still manage to be clear and persuasive), it could give you an edge over your competition.

Here are the six main information elements you should include in a proposal:

1. **Introduction.** A short overview, perhaps thanking the reader for the opportunity to be of service.
2. **Needs.** What the organization needs.
3. **Solution.** How you plan to solve the problem or supply the need. Point out any particular strong points your solution will have, and be clear about the time schedule and deadlines.

4. **Qualifications.** The experience, personnel, and skills that set your company above the competition. This is the place to promote yourself, but keep it down to a reasonable length.

5. **Cost.** How much you will charge the client and how the charges are determined. Be as specific as possible, but if there are unknown costs, be straightforward about them and make the best estimate you can. Don't be afraid to ask the client to budget 10 or 15 percent extra as a contingency against unforeseen expenses or delays. Whatever you quote, you must be prepared to live with.

6. **Other.** This might include résumés of your key people, a list of clients, letters of testimonial, a detailed time line for the completion of the project, or other supporting material.

Depending on the project, a proposal may be one or two pages long, or it may fill many pages. Keep in mind that the object of this and all writing should be to say what you need to say, as clearly as possible, in the fewest words possible.

Once you finish the rough draft, leave it alone for a period (more than twenty-four hours, if possible), then read it as if you are the other company. Ask yourself some tough questions, and be honest about the answers. For example: Does this company (us) have the experience and expertise to do the job? Will they be able to fit in with our organization and our needs? How are they different from the other companies that submitted bids? How is this proposal unique? Is it more likely to be effective that the other proposals we've received?

Abstracts

An abstract is a short synopsis of a report, article, book, or other long document. It may be a separate document, or it may be part of the longer work, in which case it is always placed at the beginning. Some people use the terms *abstract* and *summary* interchangeably, but strictly speaking a summary appears at the end of a document and doesn't serve the same purpose as an abstract. A summary reviews and sums up what has been said, but doesn't necessarily touch on the same points or cover them in the same order that an abstract would. A summary is part of the original document, but an abstract is considered a separate entity even if it happens to be printed with the original document. The purposes of an abstract are two-fold:

- To give a general overview of the main document so that a reader can decide whether or not to read the longer piece.

- To provide enough information that readers who have only a passing interest in the topic will not need to read the entire document.

Normally an abstract requires only one paragraph, although occasional documents may need two- and even three-paragraph abstracts. Only rarely should an abstract exceed 250 words.

An abstract should be objective and not offer an opinion on the original. It should also use, as far as practical, the same tone and the same type of language as the original. Abstracts fall into one of two categories: informative and descriptive (also called *indicative*). An informative abstract is a distilled version of the key ideas, procedures, results, and conclusions of the original. A descriptive abstract is shorter than the informative version and serves merely to list the contents of the original. It doesn't describe the actual results, recommendations, or conclusions.

When you write an informative abstract, don't get bogged down in explaining all the fine details of the original document. Concentrate on answering for your reader the following questions:

- What was the objective of the original work?
- What, if any, significant procedures were used?
- What were the results of any important experiments?
- What conclusion(s) did the author(s) reach?
- What implications does this work have?

To write an abstract, use a copy of the original that you can mark, highlight, and underline as much as you need to. Start by reading the first two or three paragraphs, and then read the last two or three paragraphs. This will help you get an idea of what the author thinks is important about this document. Now go back and read the entire work, underlining or highlighting key points as you come to them. Often you will think of a concise way of expressing a concept while you're reading it. If so, write it down immediately, in the margin if you like, so you don't forget it.

After you finish, go back and write down all the points you isolated. Condense them into short, clear sentences, using the author's original language *if* it's possible to do so and still be understood. Now connect these sentences with transitional words or phrases. Keep the ideas in a logical order, preserve the tone of the original piece, and be sure to refer to the five questions mentioned above.

Definitions

A basic definition has three parts—term, genus, and differentia—and follows this format:

A _____ is a _____ that _____.

Example:

> An *x-ray machine* is an *electronic camera* that *uses nondestructive radiation to penetrate soft tissue.*
>
> *X-ray machine* is the term, *electronic camera* is the genus it belongs to, and the rest of the definition lists the ways it is different from other members of that group.

Descriptions, comparisons, and analogies can all be part of a definition. There are also specialized types of definitions, the following representing the most common:

1. **Definition by illustration.** Shows an image of some kind, a drawing, photograph, or diagram, along with a brief caption.

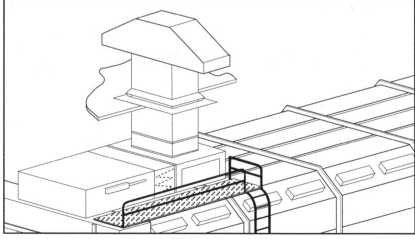

Spray booth for an assembly line

2. **Definition by stipulation.** Tells how you intend to use the term.

> *The term "fine particulate matter," as used in this report, refers to airborne particles less than 10 microns in size.*

3. **Operational definition.** A simplified form of process that explains how the defined object operates.

> *A thermometer is a sealed glass tube filled with mercury, which expands as the temperature increases.*

4. **Definition by synonym.** Consists of, or includes, a synonym, often in parentheses.

> *Rapid heart beat (tachycardia, or palpitation). A treatable condition afflicting more than two million Americans.*

Don't write definitions that are too general, too complicated, or circular.

Too general:

> *A microscope is an optical device.*

Too complicated:

> *Pin. A small (generally ranging from approximately one inch to approximately one and one-quarter inch in length) segment of extruded metal wire (often brass or tin-plated brass), having one end sharpened to a point, and with the opposite end having an integral flattened head (or in some cases, a separate globular head made from metal, glass, or plastic), and which is utilized to affix certain objects to other objects, such as a flower to a lapel.*

Circular:

> *A paper cutter is a device for cutting paper.*

Reports

The term *report* comes from Middle French and simply means a statement or account. Today it refers to a vast array of documents—some long, some short, some formal, some informal—all intended to convey information to the reader. A magazine article written by a reporter is one kind of report. A profit and loss statement prepared by an accountant is another. Reports have one or both of the following purposes:

- To inform and educate the reader
- To motivate, persuade, or inspire the reader to take action

A report is a tool for putting information that may have taken weeks to gather into a form the reader can understand in a few minutes. The format presents most or all of the following elements:

1. **Title.** Appears on the first page or on a separate front page.

2. **Table of Contents.** Is necessary only in longer reports that have several sections.

3. **Abstract.** Is not usually necessary in short reports.

4. **Executive summary.** Is similar to an informative abstract, but longer (preferably no more than one page); always placed at the beginning of the report.

5. **Opening (the "hook").** Catches the reader's attention by launching the topic in an interesting way.

6. **Overview.** Presents the problem, condition, or need.

7. **Investigation.** Tells, in greater or lesser detail, what is known about the situation; delineates possible alternatives; discusses the advantages and disadvantages of each. (Some people use the more general—and more vague—term *the body.*)

8. **Conclusion.** May be a separate section, but most often is simply the last part of the body, or investigation section.

9. **Recommendation.** Suggests the best alternative(s); points out the advantages and disadvantages (if not already covered in the investigation section).

10. **Call to action.** Stimulates the reader to take action.

11. **Summary.** Ties everything together, sums up the main points; does not introduce any new material.

12. **Appendix.** Contains supporting or reference material that was too lengthy or not important enough to be included at the relevant place in the document. (Not every report needs an appendix.)

All reporting should be objective and free of personal bias. State the facts you learned and make the logical inferences and conclusions without letting your personal feelings interfere. Give your own opinion only when you know the reader expects it, and then be sure you clearly indicate that it is your opinion. Don't editorialize or use the report as a platform for your own pet causes.

Biased:

> *Although nearly all of our technicians have mastered the process, some benighted souls, probably destined to spend their whole careers as technicians, still need almost an hour and a quarter (!!) to assemble the unit. Apparently, this is another case where the training department needs to get on the ball.*

Take out the barbed comments and stick with the facts. For example, are the terms *nearly all* and *hour and a quarter* justified by actual statistics?

Objective:

> *Although 69 percent of our technicians can assemble the unit in less than one hour, the others need longer, in one case up to an hour and seven minutes.*

As for the comment about the training department, it shouldn't even be there, unless the writer was asked to include an opinion. Then it should be stated objectively.

> *I suggest that the training department review the training program to see if there is a more effective way to teach the assembly procedure.*

Many writers think the way to be objective is simply to avoid using the pronouns *I* and *you*. The result is impersonal, but not necessarily objective. Unless your company has a style rule that prohibits writers from using these pronouns, there's no reason why you shouldn't. In many cases, you need the pronoun I because you were personally involved in the process. Otherwise, the report ends up sounding like it was written in the Victorian era.

Stuffy:

> *In the author's opinion, most technicians assemble the unit very quickly. However, it is this writer's belief that room remains for improvement.*

Clear:

> *Most technicians seem to assemble the unit very quickly, but I believe there is room for improvement.*

There is another distinct advantage in using personal pronouns: It makes the writing more readable. People like to read about people. We can read reports faster and understand issues better if the writer creates a mental image we can relate to from our own experience. The following paragraph, although it has a personal flavor, is objective without being stuffy:

> *At the request of the production manager, I tested the technicians' speed in assembling control units. Most technicians assembled the unit very quickly, but I believe there is room for improvement. Although 69 percent can assemble the unit in less than one hour, some still need up to an hour and seven minutes. I suggest that the training department review the training program to see if there is a more effective way to teach the assembly procedure.*

E-Mail

With the development of e-mail and on-line networking, many technical writers may need to learn some new conventions in writing. These practices are still developing, and refinement will undoubtedly continue for years to come, but following are a few things to keep in mind.

With some e-mail systems, there is no way to underline a word to show emphasis. You have to put asterisks around the word:

word

Further, if you want to show that something is a title, instead of underlining it, put hyphens before and after the word:

-title-

Besides the ability to transmit written documents in editable form, e-mail also offers you *talk mode* and the Internet Relay Channel (IRC). Talk mode permits you to "converse" or share information one-on-one: Other people cannot read your correspondence. With IRC, many people can talk together at once—a sort of keyboard party line.

When you use either mode, your message is seen by the other person(s) at the same time you write it (in *real time*)—a high-tech version of looking over someone's shoulder. Therefore, there's no purpose in going back and correcting spelling and grammar errors. Try to write as well as you can, but if you make an error, as long as the meaning is clear, don't bother to backspace and correct—your correspondent has already seen your gaff.

On the other hand, with regular e-mail, the reader doesn't have your message or document until you actually send it, so take the time to compose it as well as you would any business communication. Even though it may seem like you're only writing to a computer rather than another human being, the principles of good writing still apply.

To save time with IRC, talk mode, and e-mail correspondence, many people use abbreviations. The most common abbreviations are listed here in case you encounter then on-line, but don't use them yourself in formal business writing. Clarity is more important than saving another half second.

Abbreviation	Meaning
IMHO	"In my humble opinion"
BTW	"By the way"
BRB	"Be right back"
LOL	"Laughing out loud"

When you speak to someone, much of the communication is nonverbal: gestures, body language, and inflection of the voice. In writing, you have only the written word to convey the message, at least until now. Some e-mailers have developed a way of showing emotions symbolically with things called *emoticons* (emotion icons). These are made by using various punctuation marks to create the appearance of a face (an offshoot, perhaps, of the *typewriter art* that some artists experimented with in the 1970s). One drawback is that emoticons come out sideways. To understand them, you have to get in the habit of viewing them sideways. See if you can recognize the emoticons below without having to turn the book sideways. It takes only a little practice.

:-)	smiley face
:-(	frowny face; unhappy; upset
;-)	winky smiley; flirtatious
<;->	leering smile , sometimes done this way >;-)
(:o)	shock
(:/)	sarcasm

Another tool for conveying emotion is simply to put a cue word in between arrows. For example, <grin> tells the reader that you're only joking, or that you're happy about something, while <sniff> indicates sadness.

Again, these have been provided here only because you may encounter them when someone else uses them. Many business people consider these devices to be a nuisance, so avoid using them in your technical correspondence.

One final guideline for ensuring that your e-mail messages get read and convey the intended message: Don't use FULL CAPS. They are harder to read than upper and lowercase letters, and many readers consider you to be shouting and therefore rude when you use them.

Conclusion

Some people think of technical writing and other processes of logic as being like a ladder, with one step leading to the next, or like a pyramid, with each level built on the level before it. This analogy is fine but a bit limiting because it's primarily two-dimensional.

Now that you've learned the techniques and insights in the preceding chapters, a better analogy for you might be to think of written documentation as evolving like an airplane ride:

1. **File a flight plan stating where you're going and what route you'll take.** Before you write, make sure the subject isn't too broad to handle in one document. Define your objective and focus on it.

2. **Go high enough to get a general view of the territory, but don't get lost in the clouds.** Give the reader a brief overview of the subject, but don't go off on tangents.

3. **Approach the landing field.** State the main point you intend to make about the subject.

4. **Stay on course, getting gradually closer to the target.** Present the facts, data, and opinions that lead up to and support the main idea (your objective). Here's where it's critical that each idea leads logically to the next. Stay focused; this is no time to bring up related but nonessential points. Your readers must never forget the main idea of the document. Don't let them lose sight of the target you're leading them toward.

6. **Land right on target.** Make your point, stating the conclusion the data has led you to reach.

7. **Close your flight plan.** End with a brief summary of the piece.

Here's one final flying analogy: Keep your sentences and paragraphs short, using clear words your audience understands, and none of your readers will get lost in any verbal fog banks.

Appendix A

Quick-and-Easy Punctuation Guide

To do justice to any study of punctuation, you really need a comprehensive reference work such as *The Gregg Reference Manual*. However, the following pages will serve as a quick, easy-to-use guide to the most common trouble spots technical writers are likely to face.

The Comma

Commas are used:

- To separate nonessential expressions that interrupt the flow of thought. Such expressions are called nonessential because the sentence would still be complete without them.

    We selected Jan, who has an MBA, to head the team.

    It was, under the circumstances, the least we could do.

- To separate three or more items in a series, use the "serial" comma. This is the comma that comes before the words *and* or *or* in a list of three or more things.

    We'll be at your site next Monday, Tuesday, and Wednesday.

    (*Note:* Some people omit the serial comma, and usually there is no harm done. It's a good idea, however, to use it, because some sentences will be ambiguous otherwise.)

- Before the linking word (also called a *coordinating conjunction*) between two independent clauses. The most common linking words are *and, but, or,* and *nor.* Sometimes the following words are also used as coordinating conjunctions: *for, yet, so, neither,* and *whereas.*

    We will not give up, and we won't become discouraged.

    (*Note:* Don't use a comma if the subject is the same for both predicates, as in "We will not give up or become discouraged.")

The Semicolon

Use the semicolon:

- To separate two independent but closely related thoughts not connected by a coordinating conjunction.

    Your proposal is excellent; I've never read a better one.

- To separate items in a series when any item already has a comma inside it.

    We met Senator Louise Johnson; Dave Randall, the developer; Barbara Walters; and Dr. Alex De Anda, the Nobel winner.

    We'll visit sites at Rome, New York; Athens, Georgia; and Rome, Italy.

- Before transitional phrases when connecting two complete thoughts without a coordinating conjunction. The most common transitional phrases are: *as a result, consequently, for example, furthermore, hence, however, moreover, nevertheless,* and *therefore.* Transitional phrases are preceded by a semicolon and followed by a comma.

    Semicolons seem complicated; however, I think I understand.

    There are things we need to arrange; for example, buying plane tickets, packing the seismograph, and finding an experienced guide.

The Colon

Colons serve:

- To connect two independent clauses when the second clause explains the first, and there is no transitional expression or coordinating conjunction.

    He loves it when we send him to Denver: He was raised there.

    With a transitional expression, it would be:

    He loves it when we send him to Denver because he was raised there.

- As a stop before a list, or in bulleted items. The sentence will often contain one of the following expressions: *the following, as follows,* or *these.*

    The following people were honored: Julia Sanchez, Ted Sorenson, Latief Simpson, Terry George, and Hien Nguyen.

    The job has great benefits: dental insurance, a company car, and free tickets to the games.

    The sequence of events was as follows:

    1. The hand brake failed.

    2. The car began to roll.

    3. Officer Davies ran alongside and grabbed the wheel.

The Apostrophe

Use an apostrophe:

- To indicate an omitted letter.

    Myron can't make it to the going-away party.

    Her supervisor OK'd her request for a transfer.

- To form plurals of single letters or single numbers.

    The new three R's are RAM, ROM, and Rewrite.

    There are two 6's in the sign.

Note: Some style guides recommend omitting the apostrophe when forming the plurals of letters and numbers.

- To show possession (except in the possessive pronouns *hers, his, yours, ours, theirs,* and *its*).

    That is Terry's coat.

    Here is the car's best feature.

- Apostrophes are also used in technical writing to stand for feet.

    The road is 32' wide.

    The skid marks measured 78'.

Appendix A: Quick-and-Easy Punctuation Guide

More About Forming Possessives

For most words, simply add apostrophe and s.

For singular nouns ending in s, listen to how you say the word. If you make a new syllable, then add *'s*.

> *the boss's desk*

But if a new syllable would be awkward to pronounce, add only the apostrophe.

> *Achilles' heel*
>
> *for goodness' sake*
>
> *Los Angeles' freeways*

To form most simple possessive plurals, you need only add an apostrophe.

> *girls' dormitories*
>
> *parties' hosts and hostesses*

You must use *'s* to make the possessive of irregular plurals that don't end in s.

> *women's hospital*
>
> *men's clothing*
>
> *people's republic*
>
> *children's toys*

In forming a possessive with two or more nouns, add the apostrophe to the last noun if the object is possessed jointly by both.

> *John and Terry's car*
>
> *LeeAnn and Fred's house*

Add the apostrophe to each noun to show separate possession.

> *John's and Terry's cars*
>
> *LeeAnn's and Fred's coats*

In order to form plurals of people's names, usually you simply add *s*.

> *three Judys*
>
> *the two Hamiltons*

If the name ends in *s, x, ch, sh,* or *z,* add *es.*

> *Liz Marsh and all the Marshes*
>
> *the Marxes will be there*

But, if adding *es'* creates an extra syllable that is awkward to pronounce, just add the apostrophe.

> *the Hastings' house (not the Hastinges' house)*
>
> *the Hodges' car (not the Hodgeses' car)*

Descriptive or Possessive: Which Is It?

Don't add the apostrophe to a word ending in *s* if the word is descriptive.

> *teachers credit union* (tells what kind of credit union)
>
> *teachers' grade books* (tells who the books belong to)

Sometimes you can think of a word as either descriptive or possessive, and either way can be correct.

Descriptive:

teachers lounge (tells what kind: a lounge for teachers)

Possessive:

teachers' lounge (tells who the lounge belongs to: teachers)

Descriptive:

six-month leave (tells what kind of leave)

Possessive:

six-month's pay (tells that the pay belongs to that period)

It helps to ask yourself if you could rewrite the phrase using *for* or *by*. If so, it's usually safe to omit the apostrophe.

drivers handbook (a handbook for drivers)

musicians strike (a strike by musicians)

Dashes

The dash is used instead of a common, semicolon, colon, or parenthesis when you want to set apart and emphasize a thought. If you use dashes too often, this effect is lost. In some cases, the meaning may even become less clear. Use dashes rarely and only when you're sure the extra emphasis is justified.

Our new marketing director—can you believe it?—is none other than our competitor's former CEO.

There is one indispensable element in customer service— courtesy.

Parentheses

While dashes provide emphatic separation, parentheses separate without emphasis. The material inside parentheses is not optional reading, but it is of less importance than the rest of the sentence. Use parentheses for nonessential elements such as:

- **Brief explanations**

    On Monday (if the shipment arrives by then), we'll restock the shelves.

- **Dates**

    John Maynard Keynes (1883-1946) negotiated the multibillion dollar post-war loan to England.

- **References**

    The problem is the refinancing requirement (see clause 16b), which could force us to pay a higher rate.

Sometimes it's difficult to decide whether to separate words by using commas, dashes, or parentheses, as in the following examples:

Weak:

The ships sent out: the Arliss, Dallas, Seattle, and Tulsa, all burned oil, which was more costly than coal in wartime conditions.

Better:

The ships sent out—the Arliss, Dallas, Seattle, and Tulsa—all burned oil, which was more costly than coal in wartime conditions.

Appendix A: Quick-and-Easy Punctuation Guide

Or:

> *The ships sent out (the Arliss, Dallas, Seattle, and Tulsa) all burned oil, which was more costly than coal in wartime conditions.*

Avoid writing parenthetical comments inside parenthetical comments; it is very confusing. If you absolutely must have a parenthetical comment inside a parenthetical comment, the order of punctuation is this: parentheses, then brackets, then French brackets, as shown below:

([{ }])

Brackets

Brackets are used by a writer or an editor to explain or clarify something in another writer's work.

> *He [Johnson] told us that he [Smith] should phone the office immediately.*

Note: The Latin word *sic* in brackets indicates an error or an ungrammatical usage in quoted text that the editor chose to leave as it appeared in the original.

> *The Governor told the Senate committee, "Tain't [sic] the weather that's uncomfortable, it's the testimony."*

Hyphens

Hyphens are used to separate parts of words. The key to hyphenation is to put as much of the word as possible on the first line, so the reader can begin to recognize it.

Hyphens are also used to clarify compound adjectives.

He was a foreign-car dealer.

Give me a follow-up report.

Words beginning with *self* are always hyphenated.

self-pity

self-educated

Underlining Titles

Underline or italicize the titles of complete works, such as books, magazines, newspapers, movies, plays, TV shows, pamphlets, bulletins, symphonies, operas, poems, essays, lectures, sermons, and reports.

Put quotation marks around the titles of articles inside complete works, songs, and episodes of continuing TV series.

His picture was in <u>Newsweek</u>, in a piece titled "New Executives."

Use either quotation marks or italics for the names of ships, trains, aircraft, and spacecraft, as well as for foreign words borrowed but not anglicized (such as "habeas corpus" or *habeas corpus*).

Appendix B

The Gunning Fog Index

Robert Gunning devised a fairly simple formula to determine how difficult a document is to read. The formula yields the "Fog Index": the number of years of formal education a person would need in order to understand that document. Here's how it works:

Step 1

From your document, select a random sample of approximately one hundred words in complete sentences. Results are more accurate if the sample doesn't come from the beginning or end of the document.

Count the number of words in the sample.

Count the number of sentences in the sample.

Divide to find the average sentence length.

Step 2

Count the long words (three or more syllables), but don't count:

- Proper names (such as *Washington*).
- Combined words (such as *dreamworld*).
- Words with prefixes or suffixes (such as *involv-ing*).

Step 3

Add the number of long words to the number of words per sentence.

Step 4

Multiply the result by .4 to get the Fog Index.

Example

Step 1. A sample consists of 102 words and 6 sentences: 102 ÷ 6 = 17 words per sentence.

Step 2. The sample contains 13 long words.

Step 3. Add 17 + 13 = 30.

Step 4. 30 x .4 = 12.

A reader would need twelve years of education (be a high-school graduate, in other words) in order to understand this document easily.

Appendix C

Answers to Exercises

Exercise #5: Organizing Your Thoughts (pages 37 and 38)

A. Logical Sequence

9   Bring me the file folder.

7   Unlock the filing cabinet.

3   Open the right-hand drawer of the desk.

2   The antique writing desk is against the wall opposite the desk.

5   The filing cabinet key is the only silver-colored key on the key ring.

6   The filing cabinet is next to the desk.

8   Find the file folder marked "Harris Plant Floor Plan."

4   Find the key ring in the drawer.

1   The office is the second room on the right; go in.

B. From General to Specific

4   Automobiles

9   Carburetor

6   Ford

10 Carburetor idle-adjustment screw

3   Gasoline-powered vehicles

1   Transportation

5   American cars

7   Grandma's 1967 Ford T-bird

2   Vehicles with wheels

8   T-bird engine parts

Exercise #9:
Practicing Gender-Neutral Writing (page 65)

This sentence—

Whenever a technician repairs a unit in the field, he must submit Form 72/30 to the accounting department.

—could be rewritten in several different ways. Here are five.

1. Whenever a technician repairs a unit in the field, he or she must submit Form 72/30 to the accounting department.

2. Whenever technicians repair a unit in the field, they must submit Form 72/30 to the accounting department.

3. Whenever you repair a unit in the field, you must submit Form 72/30 to the accounting department.

4. Technicians who repair a unit in the field must submit Form 72/30 to the accounting department.

5. Any technician repairing a unit in the field must submit Form 72/30 to the accounting department.

Appendix C: Answers to Exercises

Exercise #10:
Stamping Out Passive Voice (page 72)

This sentence—

Occasionally, effective air pollution abatement procedures must be predicated on the total redesign of an industrial product or the retrofitting of more effective production equipment having the same features, in terms of operational characteristics, but with reduced emission levels.

—could be rewritten this way:

Reducing air pollution sometimes requires installing new equipment which works just as well, but causes less pollution. Occasionally, it may even be necessary for the manufacturer to redesign the product in order to reduce pollution.

(Nobody said you had to leave it as one sentence. But maybe your way is even better.)

Exercise #12: Improving Your Proofreading Accuracy (page 91)

1. We can't glue it back together, but maybe we can still do the test.

2. The video project requires four items: a letter of agreement, a treatment, a first draft, and the final script.

3. Although no one ever heard of him, Dr. Monty had written many scholarly articles, including "The Structure of Polymorphonuclear Leukocytes."

4. After writing his book, <u>Polymorphonuclear Leukocytes, Elvis, and the JFK Conspiracy</u>, Dr. Monty became a well-known guest on TV talk shows.

5. Jerry replied, "What I said was, 'It's well known that Dr. Monty wrote <u>Polymorphonuclear Leukocytes, Elvis, and the JFK Conspiracy</u>.' Didn't you understand me?"

6. The recommendations team—David Boric, Dr. Karen Joyce, and C. Phil Morris—was highly qualified to analyze the Copperfield project.

7. The committee's reports vetoed Karen's suggestion for these reasons: it's too costly; the time, so they said, wasn't appropriate; and it wasn't their idea, so it couldn't be any good.

8. Karen quit and started her own company; she is now a millionaire. Dr. Monty, taking a page from Geraldo's book, got his own talk show, where he interviews vampires, witches, and politicians' ex-wives.

Bibliography and Suggested Reading

Angell, David, and Brent Heslop. *The Elements of E-mail Style*. Reading, MA: Addison-Wesley, 1994.

Bjelland, Harley. *Writing Better Technical Articles*. Blue Ridge Summit, PA: TAB Books, 1990.

Booher, Dianna, and Tom H. Hill. *Writing for Technical Professionals*. New York: John Wiley & Sons, 1989.

Cain, B. Edward. *The Basics of Technical Communicating*. Washington, DC: American Chemical Society, 1988.

Mali, Paul, and Richard W. Sykes. *Writing and Word Processing for Engineers and Scientists*. New York: McGraw-Hill, 1985.

Mancuso, Joseph C. *Mastering Technical Writing*. Reading, MA: Addison-Wesley, 1990.

Miller, Casey, and Kate Swift. *The Handbook of Nonsexist Writing: For Writers, Editors and Speakers.* 2nd ed. New York: Harper & Row, 1988.

Sabin, William A., *The Gregg Reference Manual.* 7th ed. Lake Forest, IL: Glencoe, 1992.

Schramm, Wilbur, and Donald F. Roberts. *The Process and Effects of Mass Communication.* Urbana, IL: University of Illinois Press, 1971.

Weisman, Herman M. *Basic Technical Writing.* Columbus, OH: C.E. Merrill Publishing, 1985.

Young, Matt. *The Technical Writer's Handbook: Writing With Style and Clarity.* Mill Valley, CA: University Science Books, 1989.

Available From SkillPath Publications

Self-Study Sourcebooks

Climbing the Corporate Ladder: What You Need to Know and Do to Be a Promotable Person *by Barbara Pachter and Marjorie Brody*

Discovering Your Purpose *by Ivy Haley*

Mastering the Art of Communication: Your Keys to Developing a More Effective Personal Style *by Michelle Fairfield Poley*

Organized for Success! 95 Tips for Taking Control of Your Time, Your Space, and Your Life *by Nanci McGraw*

Productivity Power: 250 Great Ideas for Being More Productive *by Jim Temme*

Promoting Yourself: 50 Ways to Increase Your Prestige, Power, and Paycheck *by Marlene Caroselli, Ed.D.*

Risk-Taking: 50 Ways to Turn Risks Into Rewards *by Marlene Caroselli, Ed.D. and David Harris*

The Technical Writer's Guide *by Robert McGraw*

Total Quality Customer Service: How to Make It Your Way of Life *by Jim Temme*

Write It Right! A Guide for Clear and Correct Writing *by Richard Andersen and Helene Hinis*

Spiral Handbooks

The ABC's of Empowered Teams: Building Blocks for Success *by Mark Towers*

Assert Yourself! Developing Power-Packed Communication Skills to Make Your Points Clearly, Confidently, and Persuasively *by Lisa Contini*

Breaking the Ice: How to Improve Your On-the-Spot Communication Skills *by Deborah Shouse*

The Care and Keeping of Customers: A Treasury of Facts, Tips and Proven Techniques for Keeping Your Customers Coming BACK! *by Roy Lantz*

Dynamic Delegation: A Manager's Guide for Active Empowerment *by Mark Towers*

Every Woman's Guide to Career Success *by Denise M. Dudley*

Hiring and Firing: What Every Manager Needs to Know *by Marlene Caroselli, Ed.D. with Laura Wyeth, Ms.Ed.*

How to Be a More Effective Group Communicator: Finding Your Role and Boosting Your Confidence in Group Situations *by Deborah Shouse*

How to Deal With Difficult People *by Paul Friedman*

Learning to Laugh at Work: The Power of Humor in the Workplace *by Robert McGraw*

Making Your Mark: How to Develop a Personal Marketing Plan for Becoming More Visible and More Appreciated at Work *by Deborah Shouse*

Meetings That Work *by Marlene Caroselli, Ed.D.*

The Mentoring Advantage: How to Help Your Career Soar to New Heights *by Pam Grout*

Minding Your Business Manners: Etiquette Tips for Presenting Yourself Professionally in Every Business Situation *by Marjorie Brody and Barbara Pachter*

Misspeller's Guide *by Joel and Ruth Schroeder*

NameTags Plus: Games You Can Play When People Don't Know What to Say *by Deborah Shouse*

Networking: How to Creatively Tap Your People Resources *by Colleen Clarke*

New & Improved! 25 Ways to Be More Creative and More Effective *by Pam Grout*

Power Write! A Practical Guide to Words That Work *by Helene Hinis*

Putting Anger to Work For You! *by Ruth and Joel Schroeder*

Reinventing Your Self: 28 Strategies for Coping With Change *by Mark Towers*

Saying "No" to Negativity: How to Manage Negativity in Yourself, Your Boss, and Your Co-Workers *by Zoie Kaye*

The Supervisor's Guide: The Everyday Guide to Coordinating People and Tasks *by Jerry Brown and Denise Dudley, Ph.D.*

Taking Charge: A Personal Guide to Managing Projects and Priorities *by Michal E. Feder*

Treasure Hunt: 10 Stepping Stones to a New and More Confident You! *by Pam Grout*

A Winning Attitude: How to Develop Your Most Important Asset! *by Michelle Fairfield Poley*

For more information, call 1-800-873-7545.